Primary Geography

Pupil Book 6 Issues

Stephen Scoffham | Colin Bridge

Unit 1 Restless Earth

Lesson 1: Earthquakes and volcanoes

> What do we know about the Earth's crust?

The ground beneath our feet seems firm and solid, yet every so often earthquakes and volcanoes make it shake and crack. Earthquakes and volcanoes happen suddenly, other Earth movements happen very gradually. Sometimes fossil sea shells are found in the rocks in high mountains. This proves to scientists that these rocks were once on the seabed.

▼ Earthquakes are measured by a seismograph. The graph shows how much the Earth moved during an earthquake on a Pacific island.

Discussion

☐ What clues show that some mountains are made of rocks that were once under the sea?

☐ What are the three sections that make up the Earth?

☐ Why might volcanoes be found in lines or groups?

▼ The layers of rock on these cliffs were twisted and bent when the land was pushed up out of the sea by Earth movements.

Key words

crust	mantle
earthquake	seismograph
fossil	volcano

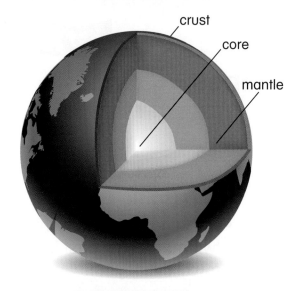

crust
core
mantle

▲ Earthquakes happen when two parts of the crust move apart or grind together. This photograph shows the San Andreas fault in California, USA, which stretches for hundreds of kilometres.

▲ When a volcano erupts, hot rocks and gases are forced to the surface. If the lava continues to flow for hundreds of years, high mountains, like Mount Ngauruhoe in New Zealand, can be built up.

Data Bank

- Between 50 and 70 volcanoes erupt each year – about one a week.
- Three-quarters of the world's volcanoes are in the 'ring of fire' around the Pacific Ocean.
- More than half the energy used in Iceland comes from hot rocks under the ground.

The Earth is made up of three different sections. The surface, or crust, is between six and 40 kilometres thick. It consists of solid rock. Beneath the crust is a section called the mantle. Parts of the mantle are so hot that the rocks have melted and flow like a sticky liquid. The core of the Earth is an even hotter ball of iron and nickel.

Mapwork

Working from an atlas or the internet, name six famous volcanoes. Add information about the date when they erupted and the country where they are found.

Investigation

Make up a diagram to show hot rocks coming to the surface in a volcano.

3

Lesson 2: Creating landscapes

Key words

erosion
glacier
landscape
limestone

What forces shape the land?

Although most rocks are very hard, they can still be cut and made into shapes. Sculptors carve blocks of stone to make statues. In a similar way natural forces shape landscapes around the world. There are five main forces.

- **Frost** Rain runs into cracks in rocks. In cold weather the water freezes breaking the rocks apart.

- **Rivers** Rivers carry particles of rock downstream creating valleys, hills and waterfalls.

- **Waves** Around the coast, waves undermine cliffs and wear away headlands.

- **Wind** Strong winds pick up particles of rock and blast them against cliffs and mountains.

- **Ice** In cold places, glaciers scrape away the rocks beneath them.

These processes all happen very slowly but over millions of years even mountain ranges can be worn down to sea level. This is called erosion.

Data Bank

- At Holderness (Yorkshire) two metres of coast is worn away every year.

- Chesil Bank (Dorset) has been built up by the sea into a shingle spit 29 km long.

▲ On the east coast of Britain, houses are slipping into the sea as waves cut away at the cliffs.

Discussion

☐ What five forces shape the land?

☐ Which of the photographs on pages 4 and 5 show erosion best?

☐ Talk about times when you have seen erosion happening in your area.

Investigation

Find out how erosion is affecting your school. Take some photographs or make drawings of wear and tear resulting from natural processes and write captions explaining what has caused it.

◄ In the desert, wind and water wear away the land. In Monument Valley, USA, pillars of hard rock have been left behind.

► Glaciers scour out deep valleys creating knife-edge ridges and jagged peaks such as the Matterhorn in Switzerland.

▼ The Niagara Falls on the St Lawrence River are moving upstream about one metre a year as the water wears away the rock.

▼ Cheddar Caves in the UK were created by water dissolving limestone rocks underground.

Lesson 3: **Rocks and soils in the UK**

Key words

clay
coal
coral reef
flint
granite
limestone
swamp

> How has the landscape of the UK formed?

The landscape that we see around us today has been shaped over millions of years. In some places the land has been pushed up into mountains. In others it has been worn away. These processes are still going on.

Volcanoes

"I am a mountain guide in Snowdonia."

▲ Snowdon in Wales.

500 million years ago most of the UK was covered by sea. Underwater volcanoes erupted forming mountains like Snowdon.

Deserts

"I am a farmer in Herefordshire."

▲ Fields of red soil, Herefordshire.

400 million years ago the land was pushed up out of the sea. The sand which rivers deposited in the desert has now become a fertile red soil.

Swamps

"I am a miner in Yorkshire."

▲ An open-cast coal mine, Yorkshire.

300 million years ago the UK was covered with swamps. Trees grew well in the hot, damp conditions. The rotting trees were pressed into seams of coal.

Coral seas

"I am a house builder in the Cotswolds."

▲ A Cotswold village, Oxfordshire.

200 million years ago the seas covered the land again. Over time sea creatures and coral reefs turned into limestone which is now used in buildings.

Rocks in the street

Rocks are valuable building materials. If you look around, you can see how people have used them in your local environment. You will probably be able to find rocks and stones which have been brought from other areas.

Rock	Way it is used
Limestone	Cut to make building blocks
Flint	Cemented together to make walls
Slate	Split into thin sheets for roof tiles
Clay	Baked in kilns to make bricks and tiles
Granite	Broken into chips to make road surfacing

brick chimney

clay tiles

flint walls

◄ These are some of the clues to look for.

road made from tar and chippings

Mapwork

Devise a trail in or around your school. Put in stopping places where different types of rock can be found including roads, pavements and flower beds.

Investigation

Build up a rock collection with labels on a display table. Ask everyone to contribute. Add things which have been made from rocks on a second table.

Discussion

How long has it taken to shape the landscape of the UK?

What processes created coal?

Why is it useful to know how rocks and soils are formed?

Summary

In this unit you have learnt:

• the way the Earth's crust moves

• about the processes which shape the landscape

• how rocks affect the character of places in the UK.

Unit 2 Drinking water

Lesson 1: Water, water everywhere

Is there enough water in the world?

There are huge quantities of water in the world. However, most of it cannot be used for drinking as it is salty seawater. Our main sources of fresh water are rivers, lakes and underground rocks.

Fresh water is essential to our lives as well as for factories and farms. Around the world, the demand for water is rising as the population increases. Also people are using more water as they buy more machines. This means that water is becoming a scarce resource. The problem is worse in crowded areas which have little rain.

Key words

borehole
pumping station
reservoir
resource
waterworks
well

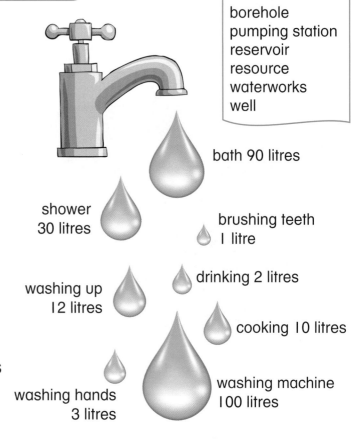

bath 90 litres

shower 30 litres

brushing teeth 1 litre

washing up 12 litres

drinking 2 litres

cooking 10 litres

washing hands 3 litres

washing machine 100 litres

▲ In the UK the average person uses around 150 litres of water a day.

Data Bank

- It takes 200 million litres of water a second to grow all the world's food.

- Seventy per cent of fresh water is in Antarctica, frozen as ice.

- The water that flows down the River Thames may have been drunk by as many as eight people before it reaches the sea.

Issues

- A dripping tap can waste 700 litres or more a month.

- Water meters have been installed in many homes in the UK to help save water.

- Around the world more people have a mobile phone than a flushing toilet.

In Britain we get more rain than we need. However, most of the rain falls in the mountains of the north and west where few people live. There the water is collected and stored in reservoirs before being pumped to other parts of the country.

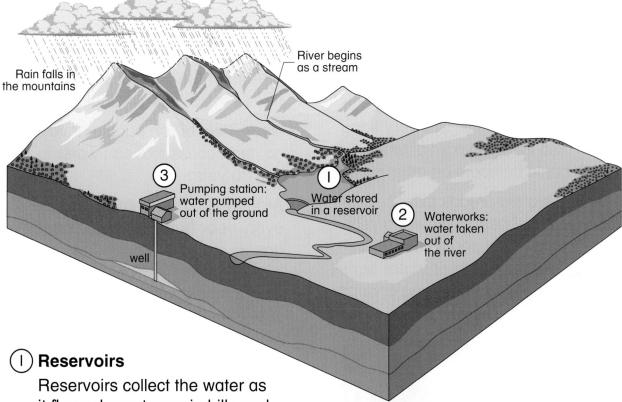

Rain falls in the mountains

River begins as a stream

③ Pumping station: water pumped out of the ground

① Water stored in a reservoir

② Waterworks: water taken out of the river

well

① **Reservoirs**

Reservoirs collect the water as it flows downstream in hills and mountains.

② **Rivers**

Sometimes water is taken straight out of rivers. It is cleaned at a waterworks and then pumped to our homes.

③ **Wells**

Wells or boreholes are built so we can reach water in rocks deep underground. The water is pumped to the surface.

▲ Kielder reservoir in Northumberland is the largest artificial reservoir in northern Europe.

Mapwork

△ Working from a map, draw a sketch map to show the region drained by your nearest river.

Investigation

○ Using data from the tap diagram create a block diagram of water use.

Lesson 2: **Water supplies**

> Why is clean water so important?

Around the world some people do not have water piped to their homes. Instead they have to collect what they need from a river or a well. Water is heavy to carry so this is a very tiring job. It also takes up a lot of time which could be spent on other work.

Polluted water spreads disease. Thousands of people die each year from illnesses caused by dirty water. Babies and young children are at risk because they are not strong enough to resist germs.

Data Bank

• A tenth of the world's population does not have clean drinking water.

• Nearly two billion people have gained access to clean water in the last 20 years.

• Around 700,000 children still die every year from diseases caused by unsafe water and poor sanitation.

Discussion

- Where do some people have to go to get their water?

- Why is clean water so important?

- If you had to fetch water from a river or well, what three things would you use it for?

▲ Women and children can spend five hours a day carrying water from the nearest well.

NORTH AMERICA

EUROPE

A S I A

AFRICA

SOUTH AMERICA

OCEANIA

▶ In some countries less than one person in four has safe drinking water.

Key

Places where many people do not have clean water

Improving water supplies

World leaders agree that everyone should have clean water. These photographs show some different ways of helping people to improve their lives.

Mapwork

Choose two photographs. Draw diagrams to show where the water comes from and how it reaches the people who drink it.

Investigation

Using an atlas and the map on page 10, name six countries where many people do not have clean drinking water.

Water pipes in Nepal

Pipes carry water from springs in the mountains to nearby villages. This gives the people a supply of clean water they can rely on.

Water tanks in India

When it rains, water from the roofs of houses runs into large storage jars. The water is used for washing and watering plants in kitchen gardens.

Dams in Bolivia

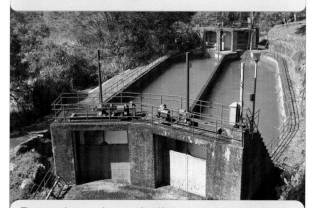

Dams have been built across streams in the hills. Water is stored during the rainy season so people have enough for the rest of the year.

Wells in Kenya

Pumps bring pure water to the surface from rocks deep under the ground. When the villagers use the hand pump, water gets sucked up to the surface.

Lesson 3: Conserving water

> Are we using water wisely?

The River Severn is the longest river in the United Kingdom. About six million people depend on it for drinking water. One of the main waterworks is at Worcester. Here water is taken out of the river, cleaned and then pumped along pipes to local houses and offices. The hospital and factories are also big water users.

Waste water flows back down drains to the sewage works. There, it is treated before being put back into the river. A few kilometres downstream there is another waterworks which takes water from the River Severn to supply other towns.

▲ Water is pumped out of the River Severn at the waterworks in Worcester.

▲ The water is used by people who live and work in the town.

Investigation

Devise a survey to discover if people use water wisely in your school.

▲ Dirty water is cleaned at the sewage works. The clean water is returned to the river which flows on towards the sea.

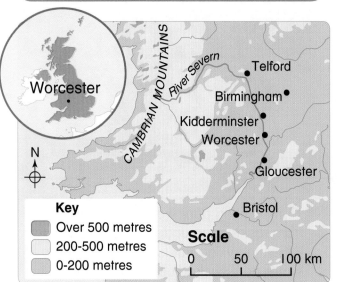

Worcester

River Severn

CAMBRIAN MOUNTAINS

Telford
Birmingham
Kidderminster
Worcester
Gloucester
Bristol

N

Key
Over 500 metres
200-500 metres
0-200 metres

Scale
0   50   100 km

Discussion

Where does the River Severn begin and finish?

Who uses water from the River Severn?

Why do you think people sometimes waste water?

A water survey

Children in two different schools decided
to investigate if they were saving water.

School A

Taps turn
themselves off
automatically.

①

Toilets have
small cisterns.

③

Plants are
watered with
rainwater which
runs off the roof
into a tub.

⑤

School B

Taps keep
running if they
are left on.

②

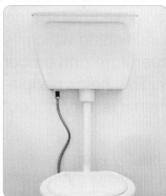

Toilets have
large cisterns.

④

Rainwater runs
straight down
the drains and
is not saved for
other uses.

⑥

Mapwork

Draw a diagram to show how
water is taken out of the River
Severn, used in different ways
and put back again.

Summary

In this unit you have learnt:

• where drinking water comes from

• that polluted water causes illness

• how people can save water.

Unit 3 Local weather

Lesson 1: The right conditions

> Why do people want to control the weather?

Key words

air conditioning
environment
humidity
temperature
ventilation

Throughout the ages, people have used their skill and intelligence to overcome extremes of hot and cold. We try to control our environment in order to feel comfortable.

Cars and houses have heating systems which operate in cold weather. Some offices have air conditioning which regulates the temperature and amount of humidity in the air. This helps people to work well even when it is very hot outside.

Discussion

☐ Why do people want to control the weather?

☐ How can special glass domes benefit holiday makers?

☐ What type of weather do you (a) like most (b) dislike most?

▼ The inside of this holiday village is kept warm, whatever the temperature outside.

14

Data Bank

- The human body works best at 37.5°C. We feel ill if our body temperature changes even by one degree.

- In a smart house the temperature and lighting can be controlled at a distance using a mobile phone.

- Air conditioning was invented around 1902 by an American, Willis Carrier.

Mapwork

Take temperature readings in different parts of your school. Record the results on a plan using three categories – warm, cold, average.

▲ Inside glass domes, plants grow well and people can wear summer clothes.

Sometimes people travel to other parts of the world because they want to experience a different climate. For example, they might want to ski or sunbathe. However, it is also possible to go to holiday centres which are protected by special glass domes. In warm weather, the dome opens to let in fresh air. In cold weather the dome closes and the heating comes on. People can relax in the heat and go swimming even if there is snow on the ground outside. This saves making expensive journeys.

Investigation

Design a survival capsule where you could keep dry, control the temperature and have good light and ventilation.

15

Lesson 2: Micro-climates

Key words

exposed
lichen
micro-climate
sheltered
slope

> How does the weather vary between places?

Places which are close together sometimes have very different weather conditions or micro-climates. The biggest contrasts are in the mountains. There are three main influences on the weather.

Wind

Hill tops and places which are exposed to the wind tend to be colder than those that are sheltered. Plants find it harder to survive because the wind damages their leaves.

Slopes

Slopes which face the sun warm up faster than surrounding areas. Slopes which face away from the sun tend to be cold and damp.

Height

The air cools 1°C for every 100-150 metres in height. This explains why high mountains have snow on them all year round.

▲ A village in the Alps.

Alps

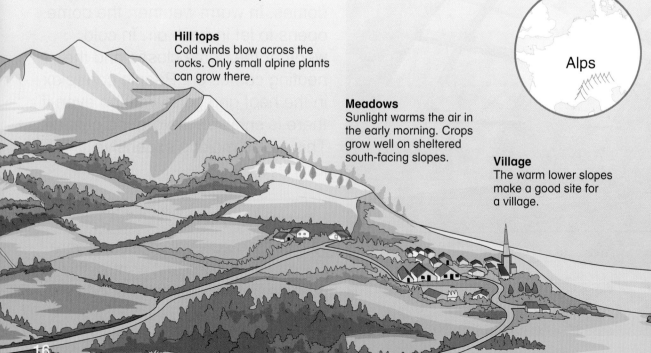

Hill tops
Cold winds blow across the rocks. Only small alpine plants can grow there.

Meadows
Sunlight warms the air in the early morning. Crops grow well on sheltered south-facing slopes.

Village
The warm lower slopes make a good site for a village.

▲ In summer, cows are taken to the upper slopes to graze on the fresh grass.

Discussion

☐ Where do weather conditions tend to vary most?

☐ What three factors combine to create local weather conditions?

☐ How do people in the Alps make use of local weather conditions?

Investigation

○ Select two photographs from travel brochures, which try to attract people by showing local weather conditions.

Data Bank

- In flat areas, like Lincolnshire, farmers plant lines of trees to shelter crops from the wind.

- In the Netherlands, tomatoes, lettuces and other crops are grown in greenhouses.

- At night, towns and cities are usually warmer than the surrounding countryside because buildings give off the heat they have absorbed during the day.

Peaks
The clear, thin air keeps the temperature low even when it is sunny. Only lichens survive on the bare rocks.

Upper slopes
Snow keeps the air cold throughout the day. People ski on the slopes.

Forest
North-facing slopes stay cool. Fir trees grow well.

Valley bottom
Early morning mists develop in the valley. Ferns and mosses thrive in the damp conditions.

Lesson 3: **Influencing the weather**

How are people affecting the weather?

Cars, factories and power stations put huge quantities of fumes into the atmosphere. People are worried that air pollution is getting so bad it is having a serious effect on the environment.

Key words

acid rain	ozone hole
carbon dioxide	pollution
global warming	smog

▲ In 2013 fumes from forest fires drifted over Singapore causing dangerous pollution.

Acid rain

The wind carries sulphur and nitrogen pollution across large parts of Europe, North America and East Asia. When the fumes fall to the ground, they kill trees, poison lakes and eat into stonework.

Ozone hole

Each winter dangerous gases build up in the air over Antarctica and the North Pole. These gases react with the sunlight in the spring. They destroy the layer of ozone which protects the Earth from harmful rays from the sun.

Smog

Traffic fumes cause smog in many cities, especially in summer months. In Britain on average two children in every class suffer from breathing problems. Environmental groups think air pollution may be the cause.

Global warming

In the last few hundred years, the amount of carbon dioxide in the atmosphere has increased by nearly a half. This is trapping heat from the sun and making temperatures rise. If sea levels rise and climate patterns change it could create worldwide disruption.

Discussion

- What is putting huge quantities of fumes into the atmosphere?
- Why does air pollution matter?
- What could people do to stop air pollution getting worse?

Finding the right site

At Hempstead School, the children bought some plants to put in the school grounds. When they looked at the labels, the children found that each plant needed different growing conditions.

The children decided to make a survey of sites around their school to find the best place for each plant. They recorded information about the amount of sunshine, wind and water on a plan of the school.

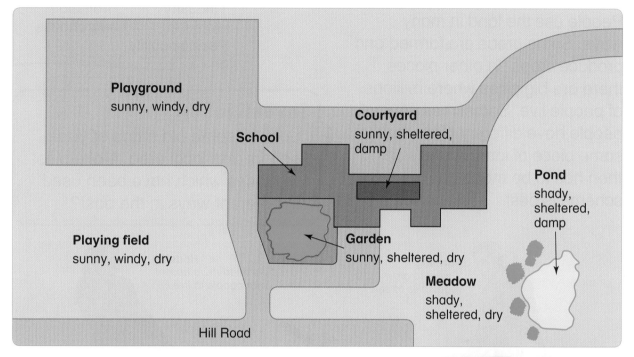

Playground
sunny, windy, dry

School

Courtyard
sunny, sheltered, damp

Pond
shady, sheltered, damp

Playing field
sunny, windy, dry

Garden
sunny, sheltered, dry

Meadow
shady, sheltered, dry

Hill Road

Mapwork

Where do you think the children from Hempstead School put each plant? Draw a plan to show where you would put the plants in your school.

Rose
Needs sunshine and water

FERN
Plant in damp shady spot

GERANIUM
grows best in dry sunny position

Investigation

Imagine you are writing a newspaper five years from now. Make up a 'good news' story about air pollution. Add pictures or drawings to go with your article.

Summary

In this unit you have learnt:

- how people can control their environment
- about local weather conditions
- about air pollution problems.

Unit 4 Planning issues

Lesson 1: Reasons for development

> Why are there conflicts over land use?

People use the land in many ways. Some areas are farmed and produce crops. In other places there are big cities where millions of people live. Sometimes several people have different plans for the same piece of land. A decision then has to be made about which scheme is best.

Investigation

Look at some old maps of your school and local area. Make a list of places which have been used in different ways in the past?

Farming
"I want to grow more vegetables to sell."

Housing
"We need more houses for people to live in."

Leisure
"We need better sports and leisure facilities."

Transport
"We must build more roads to cope with the increase in traffic."

EMPTY LAND FOR SALE

Environment
"It is essential we protect rare plants and animals."

Industry
"A new factory will bring lots of jobs to the area."

Living on an island

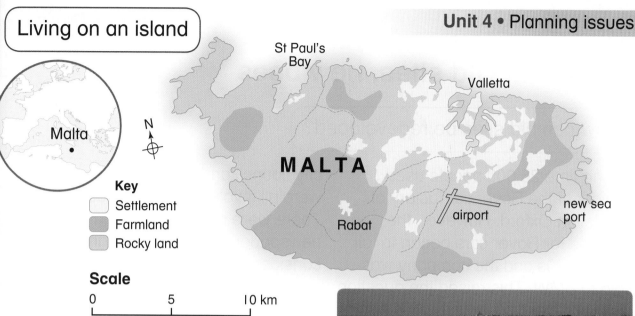

Key
- Settlement
- Farmland
- Rocky land

Scale

0   5   10 km

Malta is a small island in the Mediterranean Sea between Italy and North Africa. In the past, buildings covered only a small part of the island. Now they cover a third of the land. The new developments include:

- large, modern houses in country areas
- factories and warehouses
- a new port to encourage trade
- hotels for tourists
- greenhouses to grow vegetables
- rubbish dumps.

▲ Houses are spreading over the countryside in Malta.

The government has made strict planning laws to control development. New planning schemes have to be considered carefully.

Sustainability is a key issue. Seeing that there are good water supplies is a major challenge. Disposing of waste is another problem. In the future people will have to make very careful decisions as there is little land to spare.

Discussion

Look at the picture on page 20. Which type of land use do you think is most and least important?

Talk about a site near you which could be redeveloped. What might it be used for?

Why doesn't the government stop all new development in Malta?

Mapwork

Put a transparent centimetre square grid over the map of Malta. Work out approximately how much land is used for different purposes. Draw a bar chart of your result.

Lesson 2: **Old sites, new uses**

How can old sites be redeveloped?

At one time 25,000 people used to work at the Rover car factory at Cowley in Oxford. However, the demand for Rover cars began to fall in the 1980s and the company was sold to BMW. This raised the question of what to do next.

Local people, the car company, the city council and the county council were all involved in the discussions. There were regular reports in the local newspapers. Three main plans were considered. Eventually there was a public enquiry where everybody who was interested could give their opinion.

▲ Workers on their way home from the Rover factory in the 1950s.

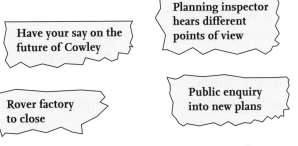

Have your say on the future of Cowley

Rover factory to close

Houses or shops for old car factory site?

Planning inspector hears different points of view

Public enquiry into new plans

Job losses at Cowley

Plan	Advantage	Disadvantage
Keep the existing factory open	Preserves jobs and keeps the factory in use	Old factory difficult to modernise and expensive to run
Close the factory and redevelop the land for housing	Helps to provide homes for the people of Oxford	Factory workers would lose their jobs
Use the land for a mixed development of offices, shops, new factories and a hotel	Creates over 4000 new jobs and provides shops for local people	Fails to create any new houses or leisure facilities

Discussion

Why did the Rover factory close?

What were the new plans?

Do you think anything was missing from these plans?

▲ Land for sale and let at the Cowley site.

▲ New offices.

Quorum research centre

Key

Hotel and offices
Building land
Shopping centre
Car factory
Roads
Trees

Oxford

Parkway Court offices

Oxford bypass

supermarket site

hotel

shopping centres

car factory

▲ New car factory.

In the end, it was agreed that the mixed development was the best option. This suited the company as they could make money from selling the site, it provided a hotel for tourists and improved the environment. Above all, it benefited local people by creating work and improving shopping facilities.

Data Bank

• Town planning dates back to Roman times.

• Land which has been built on before is known as a brownfield site.

• Land very rarely goes back to being countryside once it has been developed.

Investigation

Write sentences explaining (a) why the site needed to be redeveloped, (b) the different plans suggested, and (c) why the mixed development seemed best.

Mapwork

Imagine your school has moved to a new site and is to be redeveloped. Devise a plan of your ideas for the site.

Lesson 3: **Planning game**

> How are planning decisions made?

At Maryland Primary School, the children played a planning game. They found out how the land around them was used by making a list of the things they could see on a map and an aerial photograph. Next their teacher asked them to think what other things could be done with the school site.

> **Maryland School for Sale**
>
> Maryland School is very spacious. There are two big halls, 14 rooms and two gigantic playgrounds. This is an exciting chance to buy a very special building. It could be used for many things such as a health centre, hotel or museum.

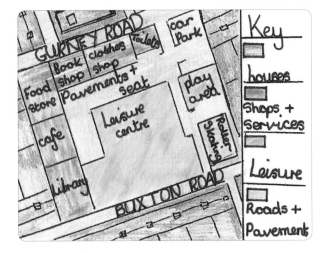

▲ The class was divided into groups and each group designed a new plan for the site. One group suggested an ice rink, another an adventure playground. The scheme for a leisure centre was very popular.

Discussion

- [ ] What can you learn from an aerial photograph which isn't shown on a map?
- [ ] What seems the main selling point for the Maryland School site?
- [ ] Why might children have different ideas to adults?

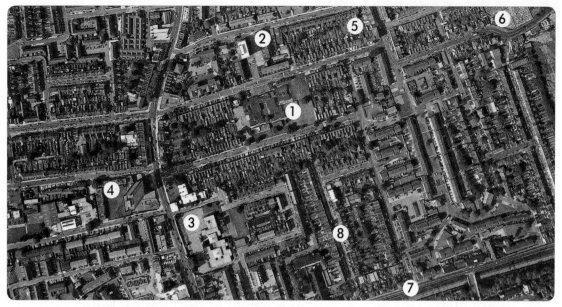

▲ An aerial photograph of the area around Maryland Primary School.

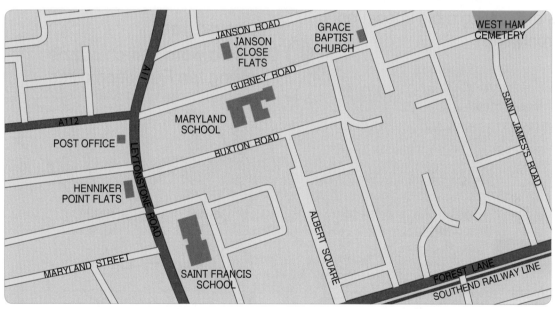

▲ A map of the same area.

Investigation

Write a 'for sale' advertisement for your school site.

Mapwork

Look at the map and photograph on page 25. Make a list of the numbered features.

Summary

In this unit you have learnt:

• that people want to use land in different ways

• how planning decisions are made

• how to obtain information from maps and aerial photographs.

Unit 5 Transport

Lesson 1: Travelling further, travelling faster

What are the opportunities for travel in the world today?

The number of people in the world is increasing. At the same time, more and more people want to travel greater distances as easily as possible. Road, rail and air transport is constantly being improved to meet these needs. However, these changes are having a big impact on the environment.

Railways

Across Europe railways are being improved with a new high-speed network. Most of the main cities are now linked together with trains that can travel at 350 km an hour.

Investigation

Devise a word search puzzle involving ten European cities.

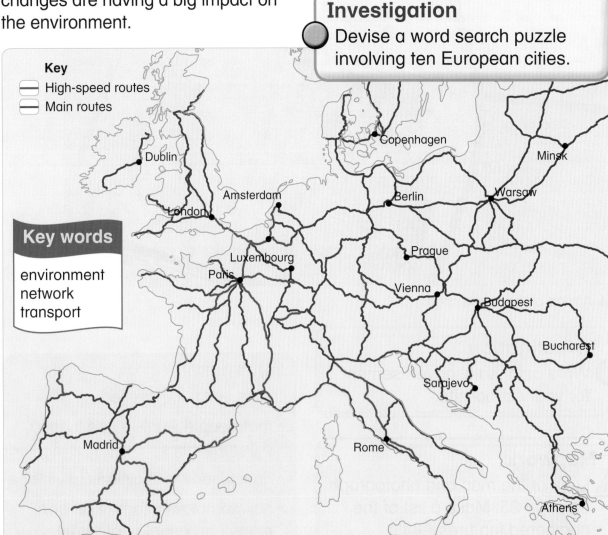

Key
- High-speed routes
- Main routes

Key words

environment
network
transport

Copenhagen
Minsk
Dublin
Amsterdam
Berlin
Warsaw
London
Prague
Luxembourg
Paris
Vienna
Budapest
Bucharest
Sarajevo
Madrid
Rome
Athens

Air travel

Since they were invented a hundred years ago, aircraft have been getting steadily larger and better. The airbus can already carry over 500 people. In future even bigger planes and larger airports may be needed to cope with demand.

Discussion

- Why do people want to travel more now than in the past?
- Why do we need different types of transport?
- What are the good and bad points about air travel?

Data Bank

- Spain has the longest network of high-speed lines in Europe.
- Five hundred million people a year travel by air.
- Luxembourg has the highest level of car ownership in the world.

Mapwork

Which places on the map are on three or more air routes?

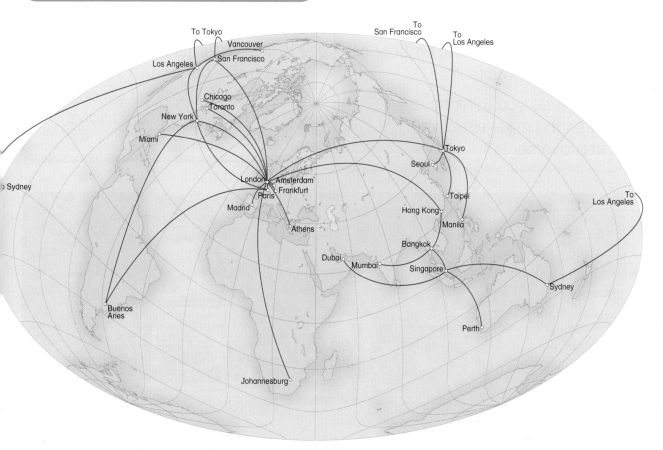

▲ Air routes link the main cities of the world.

Lesson 2: Transport problems

> Can roads cope with more traffic?

Cars are very convenient. They take us exactly where we want to go. They are comfortable to sit in and can carry lots of luggage. Today there are more and more vehicles on the roads. The amount of traffic has doubled in the last 40 years. The government has spent millions of pounds improving the roads. However, people now realise that it is impossible to keep up with the increase in traffic. As roads are improved they attract more vehicles.

Data Bank
• The first roundabout in the UK was built in Letchworth Garden City in 1903.

• Motorways are twice as safe as other roads.

• The M25 is the longest ring road in the world (195 km).

Discussion
☐ What is convenient about cars?

☐ Why do you think the amount of traffic has doubled?

☐ Which do you think have been the most effective improvements?

Investigation
○ What problems for people and traffic are shown in the photograph? Make a list and suggest how you could solve each one.

▲ Sometimes large lorries have to go through the narrow streets of old, historic towns.

Keeping traffic moving

A lot of time and money has been spent trying to keep traffic moving. This diagram shows changes over the past hundred years.

British city streets blocked with traffic.

⬇

Roundabouts improve traffic flow.

⬇

Traffic lights control road junctions.

⬇

One-way streets introduced.

⬇

Underpasses and flyovers take traffic through urban areas.

⬇

Bypasses take traffic around villages and towns.

⬇

Motorway network built to link major cities.

⬇

Roads today busier than ever.

▲ London streets in the early twentieth century.

▲ A traffic jam on the M25 motorway.

Mapwork

⚠ Make a detailed plan of a local street. Add notes about the different rules which have been devised to keep people safe and traffic moving.

Lesson 3: **Hidden costs**

Key words

acid rain
campaign
questionnaire
smog
vehicle

> How do vehicles affect people and the environment?

All over the world, cars and lorries are damaging the environment.

Resources
Vehicles are expensive to make, they consume large amounts of petrol and are difficult to dispose of when they wear out.

Air pollution
Exhaust fumes build up in the atmosphere causing smog and acid rain.

Health
Health problems such as asthma and cancer can be caused by air pollution.

Noise
Cars and lorries bring noise to the countryside and cities causing stress to people who live near busy roads.

Wildlife
Motorways and bypasses cut through the countryside damaging plant and animal habitats.

Discussion
☐ What resources do cars use?

☐ In what way might Skye be damaged by the bridge?

☐ What is the worst problem caused by traffic?

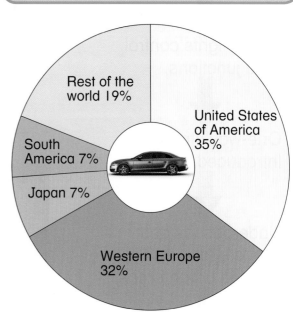

Rest of the world 19%

United States of America 35%

South America 7%

Japan 7%

Western Europe 32%

▲ There are 600 million cars in the world.

Building a new bridge
Some people argue that the bridge which now links the island of Skye to the mainland of Scotland has spoilt its character.

Skye

Finding out about local transport

At Vale View Primary School the children made a survey of traffic problems as part of a road safety campaign. First they made a list of ten problems. Then they devised a questionnaire to find out which was most serious. They asked passers-by in their local High Street for their views. The children coloured a box for each answer.

Data Bank
- Vehicles cause up to 95 per cent of air pollution in cities.
- Car exhausts contain fumes which contribute to global warming.
- Car fumes mix with water vapour to produce acid rain.

	Traffic problems questionnaire														
	1	2	3	4	5	6	7	8	9	10	11	12	13	14	15
Noise from traffic	■	■	■												
Not enough crossing places	■	■	■	■	■										
Traffic travelling too fast	■	■	■												
Roads with no pavements	■	■	■	■											
Not enough safety barriers	■	■	■	■	■										
Too many parked cars	■	■	■	■											
Shortage of cycle routes	■	■	■	■	■										
Exhaust fumes	■	■	■	■	■	■	■	■	■	■	■	■	■		
Heavy lorries	■	■	■	■	■	■									
Rush hour traffic jams	■	■	■	■	■	■	■	■	■	■	■	■			

▲ These are the results of the traffic project.

Investigation
Make a similar survey of traffic problems in the area around your school.

Summary
In this unit you have learnt:
- that traffic problems are difficult to solve
- about different schemes to control traffic
- how people can change their travel habits.

Key words

endangered
extinct
mahogany
pesticides
teak

Lesson 1: Threatened wildlife

Why are many plants and animals endangered?

All over the world wildlife is being threatened by people. Some animals are killed by accident because of pollution. Others are hunted for food or for their skins. The most serious threat comes from changes in the landscape. As cities grow larger and more land is cleared for farming, there is less space left for animals and plants.

Discussion
☐ What is the most serious threat to wildlife?

☐ How many plants and animals might be left 50 years from now?

☐ Does it matter if a plant or animal becomes extinct?

◀ In some countries, tigers are worth more dead than alive because their bones are used to make medicines.

There are probably about 30 million different plants and animals in the world today. Scientists fear that half the world's wildlife could disappear in the next 50 years. Tigers, elephants, bears, whales, crocodiles and turtles are all endangered. So too are many types of tree, flower, fish and insect.

Different plants and animals are an essential part of the world in which we live. Many medicines are obtained from plants. We eat fruit and vegetables which once grew wild. However, large numbers of plants and animals could become extinct before we can learn anything about them.

▼ The World Wide Fund for Nature (WWF) and other groups are trying to save animals and plants from extinction.

Data Bank

- Life first evolved on Earth around 3000 million years ago.

- The first fish evolved around 440 million years ago.

- Humans have lived on the Earth for the past 4 million years.

Investigation

Find out more about one threatened plant or creature. Write a short report and add pictures.

Mapwork

Draw small pictures of endangered animals such as tigers and turtles. Pin them on a large world map as a class display.

Whales
Hunted for meat and oil.
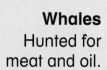

Eagles
Poisoned by pesticides.

Butterflies
Numbers declining as their habitat is destroyed.

Orchids
Dug up for house plants.

Rhinos
Killed for their horns.

Mahogany and teak
Trees cut down to make furniture.

Lesson 2: Antarctica

Key words

iceberg
satellite
treaty
wilderness
world park

> Why should Antarctica be conserved?

Antarctica is the last great wilderness on Earth. It is covered by a huge sheet of ice and is cut off from the other continents by stormy oceans. There are over 50 research bases in Antarctica where scientists study the wildlife and the weather.

The layers of ice which have built up over thousands of years provide valuable clues about the climate long ago. Scientists believe that they have found evidence that pollution from cars and factories is causing the climate to get warmer.

▼ A satellite image of the continent of Antarctica.

South Pole

Data Bank

- Three-quarters of the world's fresh water is stored in the ice and snow of Antarctica.

- In winter the sea freezes around Antarctica for nearly 100 km.

- There are no trees on Antarctica.

- In 2000, the largest ever recorded iceberg broke off the Antarctic ice shelf (295 km long).

Discussion

☐ What makes Antarctica so special?

☐ Which seven countries claim parts of Antarctica?

☐ Do you think Antarctica should be a world park?

Antarctica world park

There are valuable supplies of coal and iron ore under Antarctica. In addition, the surrounding seas are some of the best fishing grounds in the world. Some countries have tried to claim different parts of the continent for themselves. However, many people think that Antarctica is best left alone. They argue that mining will do serious damage to the environment and that fishing will make whales extinct.

In 1961, twelve countries approved a treaty agreeing:

- that Antarctica would only be used for peaceful purposes.

- that any claims to territory would be left undecided.

Now a total of 46 countries have signed up. There have also been other agreements to protect the continent. However the future of Antarctica still hangs in the balance. So far the nations of the world have co-operated with each other to save Antarctica. No one knows how long it will stay as a world park.

▲ Greenpeace campaigned for eight years to obtain a ban on mining in Antarctica.

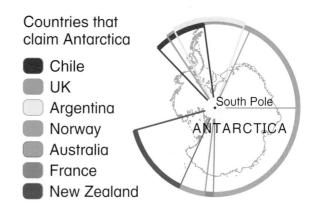

Countries that claim Antarctica

- Chile
- UK
- Argentina
- Norway
- Australia
- France
- New Zealand

Investigation

Find out about Captain Scott's journey to the South Pole. Write a page of notes with diary entries about what happened.

▼ In Antarctica, penguins survive the extreme cold by huddling together.

PROTECTED

Lesson 3: Conservation projects

Dorset

> What are people doing to conserve the environment?

In Dorset, the mild climate and light, dry soil have helped to create a special heathland habitat. Over the past 250 years, the heathlands have been split up and destroyed. By the 1980s only a few patches were left. It seemed as if even these would be lost to houses, roads and farms.

However the government realised that the heathlands were important and started to turn down development plans. The Royal Society for the Protection of Birds (RSPB) also launched a conservation campaign. Damaged areas have now been restored and isolated pockets linked by new wildlife corridors. The heathlands have been saved for future generations.

SAVED

▲ Heather and gorse are the main plants in the Dorset heathlands.

Data Bank

• Rare snakes, lizards, slow-worms, dragon flies and butterflies all live on the Dorset heaths.

• Between 1989 and 2006 the project increased the amount of good heathland by a quarter.

• Fifty landowners are now involved in helping to conserve the heathlands.

▶ The Dartford warbler can only live in a heathland habitat.

◀ Blue butterflies feed on heather and gorse.

▶ Ninety per cent of Britain's sand lizards are found in Dorset.

Discussion

☐ Why were the Dorset heathlands endangered?

☐ What made them special?

☐ What other conservation projects do you know of?

How can we keep a balanced environment?

Have you noticed that some fruit and vegetables in the supermarket have a label saying they are organic? This means that they have been grown without being sprayed by pesticides.

Jim and Pam Bennett run an organic farm in Aberdeenshire. They grow oats, potatoes, carrots, swedes and grass. They also have 40 cows and some chickens.

No pesticides or harmful chemicals are used on the farm. Jim plants a different crop in each field every year which helps to keep the soil healthy. He also uses animal dung as fertilizer.

Organic farming is hard work. Instead of using sprays, Jim does the weeding by hand. However people say the food tastes better. In addition, there are more deer, birds, badgers and butterflies in the fields.

"*We have to think of the future of the world*," Pam says. "*We want to care for the environment.*"

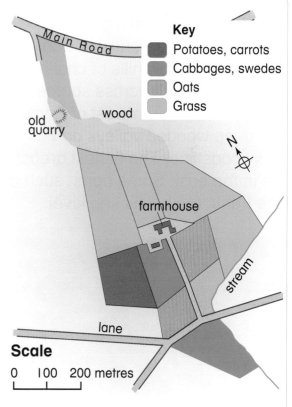

Key
- Potatoes, carrots
- Cabbages, swedes
- Oats
- Grass

Scale

0   100   200 metres

Mapwork

Make notes around a plan of your school or local area to show how it might be improved for wildlife.

Investigation

Draw picture strips to show the advantages and disadvantages of organic farming.

Summary

In this unit you have learnt:

- why wildlife is threatened
- how people and countries can co-operate to protect the environment
- how farmers can use the land without harming it.

Unit 7 England

Lesson 1: Learning about England

> ### What is England like?

England is the largest of the four countries which make up the United Kingdom. There are high hills and mountains in the Pennines and Lake District. Low hills of chalk and limestone stretch across southern England. The south west of England has deep wooded valleys and a rocky coastline. The flattest areas are in the east. Some parts, such as The Fens, are below sea level.

▲ The Lake District is an area of hills, mountains and lakes.

Discussion

- Which are the flattest parts of England?
- What do the maps tell you about each of the places listed in the 'key words' panel?
- Which area of England would you most like to visit?

Key words

Birmingham   London
Dartmoor   Pennines
The Fens   River Thames
Lake District

▲ Bulbs, like daffodils, are grown in the fertile soil of The Fens.

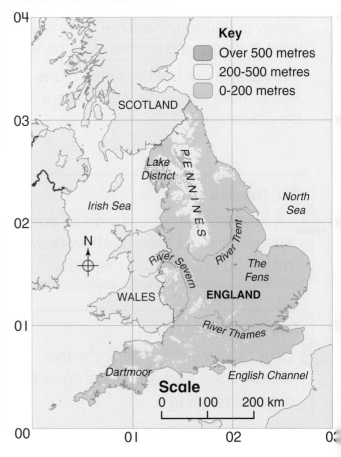

Key
- Over 500 metres
- 200-500 metres
- 0-200 metres

Scale
0   100   200 km

Blackpool Tower

Clifton Bridge, Bristol

Stonehenge, Wiltshire

Downing Street, London

Angel of the North

Weather

The Lake District is the wettest area and eastern England is the driest.

The south coast has more sunshine than any other area in England.

Rivers and landscape

The Thames, Severn and Trent are the longest rivers in England.

The highest mountain is Scafell Pike in the Lake District (978 metres).

Transport

Motorways and railways spread out from London to all parts of the country. About 50 million people use Heathrow airport (24 km west of London) every year.

Settlement

More than three-quarters of the population live in towns and cities.

Southeast England is the most crowded area.

Work

London, the West Midlands and northern England are important industrial areas.

London is a worldwide centre for banking and insurance.

Key

No of people per sq km

- Over 150
- 10-150
- Under 10

SCOTLAND

Newcastle-upon-Tyne

Irish Sea

Leeds

North Sea

Manchester

Liverpool

Sheffield

Nottingham

N

Norwich

Birmingham

WALES

Bristol

London

Southampton

English Channel

Scale

0   100   200 km

Investigation

Make a collage of ten pictures of England – two for each panel on this page.

Mapwork

Make a list of English cities arranging them in order from north to south.

Lesson 2: Finding out about Sandwich

> How has Sandwich developed?

Sandwich is a small town on the River Stour near the Kent coast. It was founded by the Anglo-Saxons over one thousand years ago who called it 'Sandwich' or 'village on the sands'.

Today Sandwich has a population of around 5000 people. The old streets and historic buildings attract tourists from the UK and abroad.

There have been many changes. A bypass takes traffic round the edge of the town. A Discovery Park is being set up on an old industrial site. There is also an energy park where there used to be a coal-fired power station.

Key words

port
ramparts

Discussion

- How long has there been a town at Sandwich?
- How has Sandwich changed?
- What do you think is special about Sandwich?

Sandwich in history

13th century Sandwich was an important port, exporting wool to Europe.

15th century The River Stour silted up and the trade stopped.

20th century Barges loaded with explosives for the First World War sailed from a secret port just north of Sandwich.

▲ Alleyways link the streets of the old town.

▲ The town wall has been turned into a walkway.

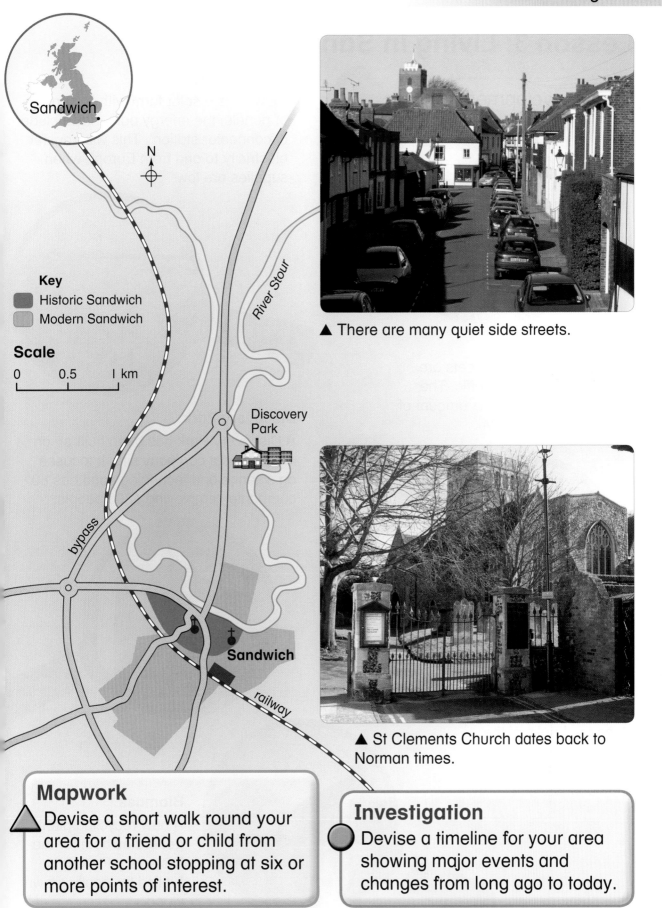

Key

Historic Sandwich

Modern Sandwich

Scale

0   0.5   l km

▲ There are many quiet side streets.

Discovery Park

River Stour

bypass

Sandwich

railway

▲ St Clements Church dates back to Norman times.

Mapwork

Devise a short walk round your area for a friend or child from another school stopping at six or more points of interest.

Investigation

Devise a timeline for your area showing major events and changes from long ago to today.

Lesson 3: Living in Sandwich

New developments

Traffic management

The narrow, winding streets are unsuitable for heavy traffic. The bypass has reduced the amount of traffic through the town.

Solar energy

As well as a solar farm with hundreds of panels, the energy park may include a 'connector station'. This will transfer electricity to and from Europe when supplies are low.

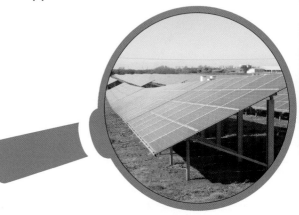

Discovery Park

A Discovery Park has been built on an old chemical company site. It focuses on science and technology and has 60 companies employing 1300 people.

New homes

New homes have been built on old meadows along the banks of the River Stour. These include both apartments and houses.

Biomass

A new recycling plant generates heat for the Discovery Park. It is fuelled by locally grown wood.

Quality of life

▼ People have different views about living in Sandwich.

I can buy all my weekly shopping in the town.

There is nothing for young people to do in the evenings.

I can walk to all the places I want to go to so I don't need a car.

I like the old buildings and quiet streets.

I get fed up with all the tourists in the summer.

I am worried by all the fumes in the air.

Living in Sandwich

Shopping

Do the shops sell most of the things you need?   ☐ Yes   ☐ No

Transport

Is it easy to walk from place to place?   ☐ Yes   ☐ No

Leisure

Are there enough facilities for young people?   ☐ Yes   ☐ No

Are there enough facilities for older people?   ☐ Yes   ☐ No

Character

Do you find the town attractive?   ☐ Yes   ☐ No

Which features do you like best?....................
....

Environment

Is the town affected by any problems?   ☐ Yes   ☐ No

List the worst problems
....
....
....

Discussion

☐ What do the people in the drawing like and dislike about living in Sandwich?

☐ Do you think all the changes are good for the town?

☐ Would you like to live in Sandwich?

Investigation

◯ Using questions from the 'Living in Sandwich' survey, find out what people think about living in your area.

Key words

character
leisure
sewage

solar panels
technology

Summary

In this unit you have learnt:

• about the physical and human geography of England

• how photographs, maps and words can give you information about a place

• how to investigate the quality of life.

Unit 8 Europe

Lesson 1: Introducing Europe

> What are the regions of Europe?

Europe is about 4000 kilometres from north to south. It has a wide variety of landscapes. These include the vast forests of Russia and high mountain ranges, like the Alps.

Discussion

- What is the landscape for the three cities marked on the map?

- If you travelled across Europe from north to south what differences would you see?

- How would you describe your area of Europe in just a few sentences?

Data Bank

- Europe is the only continent without deserts.

- Europe has a remarkably long coastline – approximately 66,000 km.

- Finland is 75 per cent forested and has the most trees in Europe.

Key words

fjord
grasslands
Mediterranean
tundra

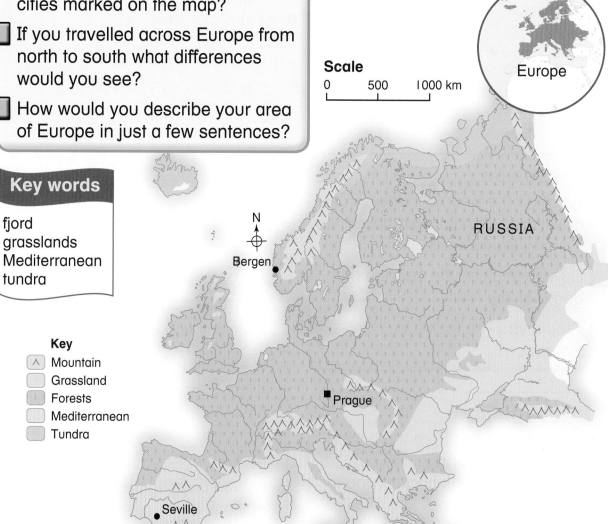

Scale

0   500   1000 km

Europe

N

Bergen

RUSSIA

Prague

Seville

Key
- ∧ Mountain
- Grassland
- Forests
- Mediterranean
- Tundra

Scandinavia

My name is Erika.

I live in Bergen in the west of Norway.

My father is the captain of a ferry boat. He takes the ferry along the coast stopping at villages on the fjords to deliver mail and other goods. People also travel on the ferries.

Central Europe

My name is Hania.

I live in Prague, the capital city of the Czechia.

My parents work in a factory which makes CDs and cassettes. At weekends we often go out into the countryside to walk in the woods and hills.

Mediterranean lands

My name is Carlos.

I live in Seville in southern Spain.

My mother works on a fruit farm. The long, hot summers and wet winters are good for growing oranges and lemons.

Mapwork

Using an atlas find a country that contains three or more of the environments shown on the map on page 44.

Investigation

Name the countries you might pass through on a journey from Seville to Bergen.

Lesson 2: The European Union

Key words

agriculture
aid
currency
tax
trade

> How can countries work together?

After the Second World War (1939-1945) much of Europe lay in ruins and had to be rebuilt. People decided that if they worked together it would help to keep the peace.

In 1957, France, Germany, Italy, Belgium and the Netherlands agreed on a scheme to develop farming and industry and increase wages. This was the start of the European Union (EU).

Discussion

☐ Why was the EU formed?

☐ How does the EU improve people's lives?

☐ Why might a country want to stay out of the EU?

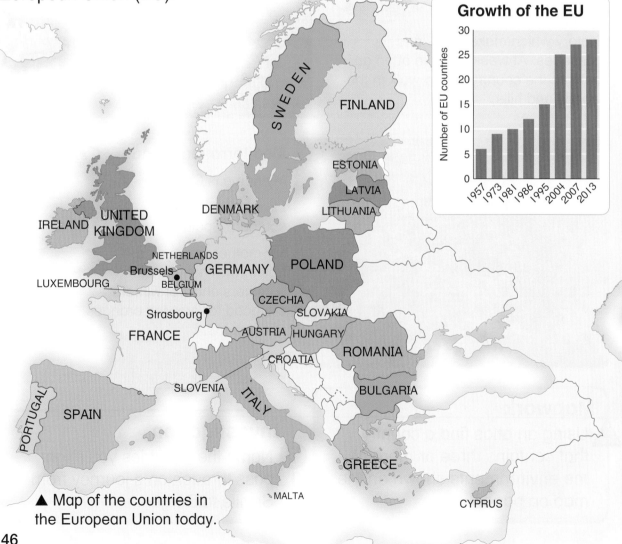

Growth of the EU

▲ Map of the countries in the European Union today.

The European Union in action

The European Union works together to improve people's lives.

Trade

All the countries in the EU co-operate over trade. Producers do not have to pay a tax when they take their goods into another country.

Joint projects

Different countries in the EU work together on big, expensive projects. This includes building aeroplanes and scientific research.

Farming

The Common Agricultural Policy (CAP) tries to make sure that people receive food at reasonable prices and that farmers are properly paid for the goods they produce.

Working conditions

Within the EU, people can work in the country of their choice without having to have special permission. There are also laws about hours of work, safety and levels of pay.

Fishing

So many fish have been caught around Europe that stocks are running low. The EU has passed laws about when fishing boats can go out to sea, where to fish and the type of nets they can use.

Environment

Many countries in the EU suffer from the same pollution problems. Laws have been agreed to protect the environment and reduce pollution levels.

Data Bank

- The EU has a population of over 500 million people.
- Each year 80,000 ships call at EU ports.
- The EU is the world's biggest aid donor.

▲ These are some of the improvements that the European Union has achieved since it was formed over 50 years ago.

Mapwork

▲ Make an alphabetical list of countries in the EU.

Investigation

What do you think are the three best reasons for a country to join the EU? Write a sentence giving reasons for each choice.

Lesson 3: Celebrating Europe

> What is special about Europe?

Last year, the children at St Mark's School organised a special European week. They searched the internet and looked in books to find out about different countries in Europe. People who had worked in Europe came into school to talk about their jobs and life abroad. Some of the children found out how to write sentences in foreign languages. Others looked at famous paintings and listened to folk music from other countries.

Data Bank

- Europe is named after a princess, Europa.
- The Vatican in Rome is the smallest country in Europe.
- Europe and North America are moving a few centimetres apart each year.

Bonjour

French

God Morgen

Danish

Guten Morgen

German

¡Buenos dias!

Spanish

▼ More than 25 countries are shown in this satellite image of Europe. How many can you name?

▲ Different ways people say 'good morning' in Europe.

EUROPE

▲ In 1963 a volcano erupted under the sea near Iceland creating the island of Surtsey.

▲ Find out about some of the different creatures which live in Europe for a class scrapbook.

▲ The Eiffel Tower in Paris is one of the most famous landmarks in the world. Try making your own tower model.

▲ Choose one of the countries or capitals on the River Danube. See how many different words you can make from the letters in its name.

Investigation

Make up a quiz with 12 questions about Europe for other children in your class.

Mapwork

Photocopy a map of Europe or make one of your own. Now cut it into around ten pieces. Challenge another child in your class to put it together again.

▲ The Acropolis in Athens is nearly 2500 years old. Make your own 'wonders of Europe' PowerPoint giving reasons for your choice.

Summary

In this unit you have learnt:

• about the different countries and landscapes of Europe

• why the European Union was formed

• what makes Europe special.

Unit 9 South America

The Amazon

Lesson 1: Learning about the Amazon

What is the Amazon like?

The River Amazon is over 6400 km long. It rises in the Andes and flows into the Atlantic Ocean. The Amazon has many tributaries flowing into it. The river basin covers a large part of South America.

The Amazon is on the Equator where the climate is very wet and warm. Over many thousands of years a vast variety of plants, animals and insects has developed. Around ten per cent of the world's living species are found in Amazonia. There are also communities of native Indians. Some of these people are living lives which are untouched by modern life.

Data Bank
- Amazonia is nearly the same size as Europe.
- There are about 400 tribes in the Amazon, each with its own language, culture and land.
- About three-quarters of the food we eat originally came from the rainforests including rice, bananas, potatoes and tomatoes.

Key words

loggers	species
native Indians	teak
rainforest	tributaries
river basin	

▲ It has taken millions of years for the Amazon rainforest to evolve.

Why is the rainforest being cleared?

Farming

Soy bean farming now covers 8 million hectares of Amazonia. Soy is used as food for both humans and cattle. As land is used to grow soy beans people are driven into the forest to clear new areas.

Cattle ranching

More and more people want to eat meat. Amazonia has a good climate for cattle ranching. This has encouraged people to clear the rainforest to make grassland for animals.

Logging

World demand for wood and paper has encouraged people to cut down trees. Sometimes loggers just remove special trees like teak but this also destroys the forest.

Roads and mines

Roads and highways are opening up more and more remote areas. In some places mines have polluted the land. In other places dams have flooded vast areas.

Mapwork

Using an atlas make a list of countries around the Amazon river basin.

Investigation

Make up your own fact file about the Amazon.

Lesson 2: Using the rainforest

What is it like to live in the rainforest?

Key words

avocados
latex
manioc
sap
thatch

"My name is José. I live in the rainforest in Brazil where my father is a rubber tapper. He collects the sap from rubber trees in the forest. Sometimes I go with him. There are many families like us in Brazil. The forest provides us with food, the materials to build houses and most of the other things we need. We use the trees without damaging them."

Discussion

☐ What is a rubber tapper?

☐ What do José and his family get from the rainforest?

☐ What could change his way of life?

We live in a wooden house

We grow rice, maize and manioc (which is like potatoes) in a forest clearing

When my father comes to a rubber tree, he cuts the bark and collects the sap, called latex in a cup

In the evening, my father lights a fire. The smoke turns the latex into rubber

We see snakes, monkeys and other animals in the forest

Sometimes my father shoots a bird or an animal for dinner

My mother knows all the plants we can use as medicines

We collect coconuts, avocados or other fruit from trees along the trail

Palm trees provide the thatch for the house roof

52

Why is the rainforest so important?

As the rainforest is cleared people are realising they are losing a unique habitat. Once it is lost it will be lost forever.

> *"Today everyone wants to make money out of the Amazon, and we are scared. Scared by the burning that is taking place, by the destruction that is taking place, by the pollution. I speak as a person who has lived in the forest all his life."*
>
> Leader of the Kayapo Indians

Investigation

Write a few sentences explaining
(a) how the trees protect the soil
(b) what happens when the forest is cleared.

Issues

- Many plants and creatures in the Amazon are becoming extinct even before they have been studied by scientists.

- As the trees are cut down the people who live in the Amazon are losing their homes.

- The trees in the Amazon put water into the air from their leaves. If too many are cut down it could cause changes in the climate around the world.

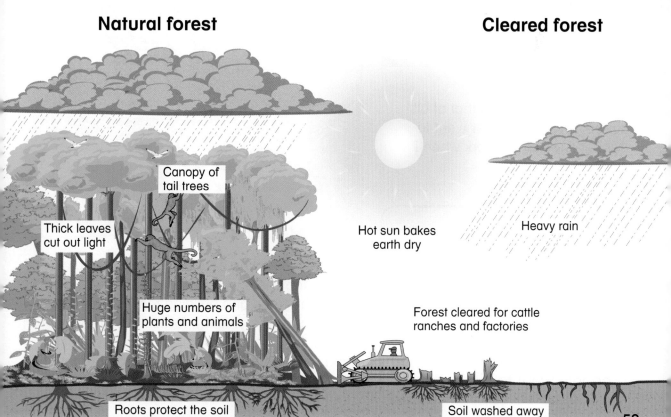

Natural forest

Canopy of tall trees

Thick leaves cut out light

Huge numbers of plants and animals

Roots protect the soil

Cleared forest

Hot sun bakes earth dry

Heavy rain

Forest cleared for cattle ranches and factories

Soil washed away

53

Lesson 3: **Saving the Amazon**

Key words

cattle   reserves
plantation   rubber
ranch   tapper

> What was Chico Mendes trying to do?

Chico Mendes was a rubber tapper. In 1988 he was murdered on the steps of his home in Brazil because he tried to save the rainforests. Thousands of people from all over the world came to his funeral. This is his life story.

1962

② When I was 18 I was taught to read and write by a rubber tapper who lived alone in the forest. We used to listen to the radio to find out what was happening in the world. Every night we talked about what we had heard.

1944

① My life began like that of all rubber tappers. I never went to school and started work when I was nine years old.

Discussion

☐ How did Chico Mendes learn about the world?

☐ What did he want the government to do?

☐ Why do you think so many people went to his funeral?

1968

SINDICATO

1974

③ I realised we needed to save the rainforest. The landowners were busy cutting and burning down the trees and building ranches for cattle. This was destroying the forest. In 1974 I joined the Rubber Workers Union. We met the leaders of the local American Indian groups and they agreed to help as well.

1985

④ We asked the government to set up reserves. We wanted people to find ways of using the forest without damaging it.

1988

⑥ After this I knew my life was in danger. Da Silva hired a gang of gunmen. However, even if I am killed the protest will go on.

1987

⑤ In 1987, the plantation where I worked was sold to a man called Da Silva. He tried to drive us off the land so he could build a ranch. We stood firm and the government made it into a reserve.

Data Bank

- In the rainy season the Amazon can be as much as 140 km wide.

- There are no bridges across the Amazon.

- Thirty-six million hectares of forest have now been protected.

Investigation

Draw a timeline showing Chico Mendes' life.

Summary

In this unit you have learnt about:

- why the Amazon rainforest is so special

- how it is threatened

- how it can be protected.

Mapwork

Write a letter that includes a map asking the government to make more reserves.

Unit 10 Asia

Lesson 1: Southeast Asia

What is Southeast Asia like?

Key words

natural resources
palm oil
peninsula
tropics
typhoons

Southeast Asia consists of a mixture of islands, peninsulas and coastal areas. It lies within the tropics and was once heavily forested. Today great cities have sprung up across the region. This has made it, along with China, one of the power houses of the modern world.

Indonesia is the largest country in Southeast Asia. It has many volcanoes and high mountains. Further north, the Philippines is another island nation. Here typhoons are sometimes a hazard.

Most of Southeast Asia was ruled by Europeans but has forged ahead since independence. Factories now produce cars, clothes and electrical goods. Palm oil is a key crop. Oil, tin and gold remain important natural resources. The development was so fast the countries were nicknamed 'tiger economies'.

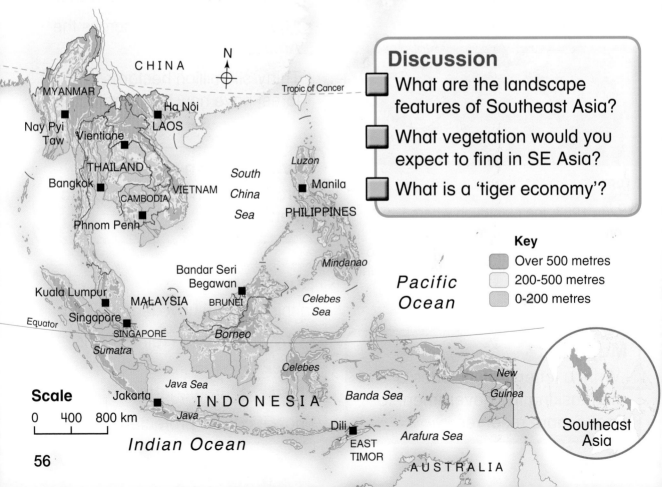

Discussion

- What are the landscape features of Southeast Asia?
- What vegetation would you expect to find in SE Asia?
- What is a 'tiger economy'?

Key
- Over 500 metres
- 200-500 metres
- 0-200 metres

Scale
0   400   800 km

Southeast Asia

▲ Crater lake, Indonesia.

▲ Palm oil plantation.

▼ Kuala Lumpur skyline.

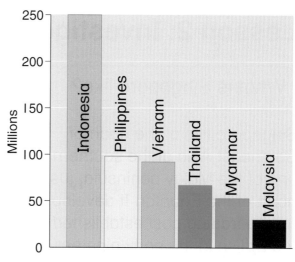

▲ SE Asia has roughly the same population as Europe.

Mapwork
Mark some of the largest cities on your own map of SE Asia.

Investigation
Devise a fact file for one SE Asian country giving details of the landscape, cities, products and environment.

Lesson 2: Investigating Singapore

Singapore

What is Singapore like?

Singapore lies at the heart of Southeast Asia at the southern end of the Malay peninsula, just north of the Equator. It developed from a trading post established by the British in the early nineteenth century. Singapore grew rapidly under British rule and quickly came to dominate the sea route between the Pacific and Indian Oceans. Today Singapore is a vast modern city with a population approaching 5 million people and a world centre for trade and finance.

Discussion

☐ How did Singapore develop?

☐ What is Singapore like today?

☐ What makes Singapore special?

Nature Reserve

The rainforest which once covered Singapore is preserved in the Bukit Timah Nature Reserve.

New towns

There are 22 new towns in Singapore.

Bridge

A busy causeway links Singapore to Malaysia, providing a route for people, traffic and water pipes.

Docks

Singapore docks handle 32 million containers a year and there are plans to double this number.

Industry

There are over 70 oil and chemical companies on Jurong Island.

Reclaimed land

As land is reclaimed Singapore island is growing larger and changing in shape.

City centre

Singapore is one of the world's largest financial centres.

Scale

0   3 km

Johor Bahru

MALAYSIA

Selat Johor

Selat Johor

Pulau Ubin

PUNGGOL

Serangoon Harbour

SINGAPORE

TAMPINES

TANGLIN

QUEENSTOWN

Strait of Singapore

Selat Jurong

Jurong Island

Selat Pandan

Pulau Busing

Pulau Hantu

Pulau Bukum

Sentosa

Pulau Sakijang Bendera

Pulau Sakijang Pelepah

▲ Offices and tall buildings tower above historic warehouses.

Climate

Singapore has a tropical climate with heavy rain throughout the year.

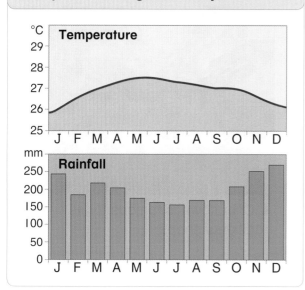

°C
Temperature
29
28
27
26
25
J F M A M J J A S O N D

mm
Rainfall
250
200
150
100
50
0
J F M A M J J A S O N D

Airport

With over 50 million passengers a year Changi Airport is one of the busiest in the world.

▲ The plants and glass domes in the Gardens by the Bay attract thousands of visitors.

Data Bank

• Singapore is the smallest country in Asia.

• Singapore is the most densely populated country in the world apart from Monaco.

• The last wild tiger was shot in Singapore less than a century ago.

Investigation

Devise an advertisement encouraging a business to set up in Singapore.

Mapwork

Working from an atlas, name some other places which, like Singapore, are close to the Equator.

Lesson 3: **A Singapore family**

What is it like to live in Singapore?

The Chow family moved to Singapore from southern China around 70 years ago. They came to escape war and arrived with only a few possessions. However the family soon found a new life. Mr Chow set up a shop as a tailor in the Chinatown area of Singapore.

Mr Chow and his wife soon had children. One of Mr Chow's sons found work in the shipping industry. He was promoted to take charge of work building the rigs and platforms which are used at sea.

▲ Singapore earns a lot of money providing services for ships.

Mr Chow has now died but his grandchildren and relatives still live in Singapore. Many of them have been to university. One of his great nieces, Jessica, works for the Singapore port authority planning new developments.

▶ Jessica Chow with her family at her graduation from university.

Discussion

☐ What job did each member of the Chow family do?

☐ How do these jobs link with the history of Singapore?

☐ What do you think is the most important issue for Singapore in the future?

▲ Chinatown is famous for its traditional shops.

Key words

community	reclaimed land
mangrove swamp	self-sufficient

Planning for the future

The population of Singapore could reach 7 million by 2030 and there will be increasing numbers of old people. The government is making careful plans for the future.

Tampins new town

Tampins is built on land which was once a forest and mangrove swamp. Groups of flats are arranged in groups around a central area with shops and meeting places. There are factories and places to work round the edge of the town and frequent automatic trains into the city centre. There are plans for more self-sufficient towns like Tampins.

▲ Each community has its own shops, schools, health centres and parks.

Investigation

Make a survey and write a report about how water could be saved or recycled in your home or school.

Drinking water

There are plans to make Singapore self-sufficient in water. A dam has been built across the harbour to catch the rain that falls on the city centre and surrounding areas. Desalination plants have also been built to remove salt from sea water.

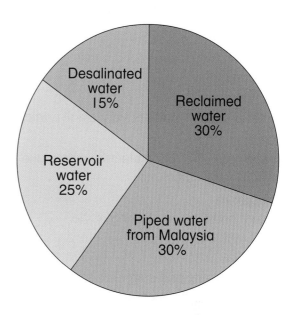

Mapwork

Devise a plan of how you could arrange six blocks of flats around a precinct. Remember to include roads, gardens, shops and meeting places.

Summary

In this unit you have learnt:

• how Southeast Asia is changing

• different aspects of Singapore

• how Singapore is planning for the future.

Glossary

Acid rain
Rain which has been polluted by fumes which damages the environment and affects people's health.

Aid
Money and/or goods which are given to help people in need.

Bypass
A route such as a road round a town which goes round an obstacle.

Clay
A water-proof soil made of very small particles.

Council
A group of people (often elected) who make decisions for a community.

Earthquake
A violent and sudden shaking of the Earth's surface.

Extinction
Species of plants and animals become extinct when they have all died out.

Flint
Layers of very hard, knobbly stones which build up in chalk and limestone.

Glacier
A thick sheet of ice which flows very slowly down slopes towards the sea.

Granite
A very hard, pink or grey, volcanic rock which is used in buildings and roads.

Greenhouse
A special building with lots of glass where plants can benefit from light and warmth.

Heathland
An area of open land covered by shrubs and heather.

Iceberg
A large piece of ice which has broken off a glacier or ice sheet to float in the sea.

Leisure centre
Building, or set of buildings, where people can go for sports, entertainment and to relax.

Limestone
A rock which forms in shallow seas from the remains of countless sea creatures.

Network
Lots of connections which join together to form a web.

Peninsula
A narrow strip of land which extends into the sea.

Public enquiry
Special meeting where people can discuss a problem in front of an inspector.

River basin
The area which is drained by a river and its tributaries.

Spring
An opening where underground water comes to the surface.

Teak
A hard and very valuable wood that comes from the rainforest.

Treaty
An agreement between different countries.

Tropics
Parts of the world where the sun is directly overhead at least once a year.

Typhoon
A very violent tropical storm which brings gales and flooding.

Volcano
An opening in the Earth's crust where red hot, underground rocks and gas break to the surface.

Warehouse
A large building used to store (and sometime sell) goods.

Primary Geography Pupil Book 6
Published by Collins
An imprint of HarperCollins Publishers
Westerhill Road
Bishopbriggs
Glasgow G64 2QT
www.harpercollins.co.uk

First edition 2014

ISBN 978-0-00-756362-3

10

Printed and bound by Martins the Printers

Most of the mapping in this publication is generated from Collins Bartholomew digital databases. Collins Bartholomew,
the UK's leading independent geographical information supplier, can provide a digital, custom, and premium mapping
service to a variety of markets. For further information: Tel: +44 (0)208 307 4515,
e-mail: collinsbartholomew@harpercollins.co.uk or visit www.collinsbartholomew.com

Acknowledgements
Additional original input by Terry Jewson
Cover designs Steve Evans illustration and design
Illustrations by Jouve Pvt Ltd pp 9, 16, 17, 20, 43, 52, 53, 54, 55

Photo credits:
(t = top b = bottom l = left r = right c = centre
© Aleksandar Todorovic/Shutterstock.com p11tl; © Anton_Ivanov/Shutterstock.com p11br;
© British Motor Industry Heritage Trust p22; © cestpin/Flickr.com p11tr;
© ChameleonsEye/Shutterstock.com p35tr; © duluoz cats/Flickr.com p23c;
© gnomeandi/Shutterstock.com p10; © Hinochika/Shutterstock.com p18;
© Irina Ovchinnikova/Shutterstock.com p11br; © joyfull/Shutterstock.com p59tr;
© Le Jhe/Flickr.com p49tl; © Mick Lobb/Geograph.org.uk p28b;
© Maryna Patzen/Shutterstock.com p16; © Miranda Smith Productions Inc./Wikicommons.com p54tl;
© net_efekt/Flickr.com p23a; © Pavel L Photo and Video/Shutterstock.com p14;
© pcruciatti/Shutterstock.com p39; © Ritu Manoj Jethani/Shutterstock.com p60tr;
© satguru/Flickr.com p23b; © Soren Egeberg Photography/Shutterstock.com p57b;
© Stephen Scoffham p12t, p12b, p24, p40, p41, p42, p60br; © Tela Chhe/Flickr.com p15;
© The Print Collector/Alamy.com p29; © zentilia/Shutterstock.com p30c

All other images from www.shutterstock.com

www.collins.co.uk/primarygeography